# 图解笔记
## DIAGRAMNOTES
### ——建筑设计概念手记
ARCHITECTURE DESIGN CONCEPT NOTES

何东明　编著

广州·北京·上海·西安

图书在版编目（CIP）数据

图解笔记：建筑设计概念手记/何东明编著.——广州：世界图书出版广东有限公司，2017.8
 ISBN 978-7-5192-3687-8

Ⅰ.①图… Ⅱ.①何… Ⅲ.①建筑设计—中国—图集 Ⅳ.①TU206

中国版本图书馆CIP数据核字（2017）第220311号

| | |
|---|---|
| 书　　名 | 图解笔记——建筑设计概念手记<br>TUJIE BIJI —— JIANZHU SHEJI GAINIAN SHOUJI |
| 编 著 者 | 何东明 |
| 责任编辑 | 刘文婷 |
| 装帧设计 | 汤　丽 |
| 出版发行 | 世界图书出版广东有限公司 |
| 地　　址 | 广州市海珠区新港西路大江冲25号 |
| 邮　　编 | 510300 |
| 电　　话 | （020）84459702 |
| 网　　址 | http://www.gdst.com.cn/ |
| 邮　　箱 | wpc_gdst@163.com |
| 经　　销 | 新华书店 |
| 印　　刷 | 虎彩印艺股份有限公司 |
| 开　　本 | 787mm×1092mm　1/16 |
| 印　　张 | 7 |
| 字　　数 | 115千字 |
| 版　　次 | 2017年8月第1版　2019年8月第2次印刷 |
| 国际书号 | ISBN 978-7-5192-3687-8 |
| 定　　价 | 88.00元 |

版权所有，翻印必究
（如有印装错误，请与出版社联系）

# 前言 | PREFACES

与其他艺术相比，建筑图（drawing）总处于某种"边缘情形"。　　　　　　　　　　　　　　——本亚明

图解是非物质的、无形的、比符号语言稍广博些的抽象机器（Abstract machine）。　　　——德勒兹

当代建筑中的"图解"是一种抽象，一种不同于"类型"的抽象："类型"经常将事物还原为常规，而"图解"则在对传统的重复中产生创新。　　　　　　　　　　　　　　　　　　　　　　　　——彼得·艾森曼

图解在设计中会把自身定位在真实与虚拟之间，置于建筑中各种交互作用的最前列。　　——斯坦·艾伦

    图解作为语言学的范型，能独立生成与建构，因此这也成为了埃森曼关于建筑本体探讨的触点，即图解使建筑学独立于其他学科形成"自治"的体系。本书的初衷未曾想论述图解与建筑本体的关系，而是希望通过归纳图解作为空间建构的操作方法，以揭示建筑概念发生的机制。

    对于操作性图解来说，伯纳德·屈米（Bernard Tschumi）将其划分为：概念图解，转录，变形序列，可互换量图四个部分。基于这四种类型的分析与归纳，张琪琳、韩冬青在《图解——期待未知》一文中，将操作性图解分为转译图解、转形图解、变形图解三个方面，笔者认为后者分类更为简易明确。本书借鉴了以上关于转形图解部分的定义：通过对纯粹几何图形的变形，得到新形式或新逻辑的操作图解；通过这个线索展开，我们汇编相关性作品，以厘清其中的变形逻辑，形成一本简明、易读、互动的"口袋书"，藉此成为设计师手头的概念笔记。

    图解蕴含着观念性、操作性、交互性三方面的机制。事实上，克里尔兄弟的类型学、科林·罗透明性研究、海杜克的"九宫格"解析等，既是空间建构观念性思考，也是关于空间生成的操作性图解。本书着眼于图解的操作性，并将空间建构机制归纳为内在性和外在性图解两个方面，其中：一方面是内在性图解，即基于原型的建构"动机"——线、面、体自身建构的机制、另一方面是外在性图解，即基于外部场地，环境对空间生成的建构"动机"——以场地要素的介入、边界的关联等形成的建构机制。尽管如此，这种简单主观的划分，不免落入"类型"化的窠臼，事实上，空间的操作性远大于其结果性，我们只有以一种开放的视角获取不同空间操作线索，才能在设计源头厘清其矛盾性和结构性，才能开启更多空间建构思路，使设计回归本体。由于编写仓促，难免疏漏与错误，望读者批评指正。

何东明

2016.12

# How to use the book?

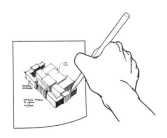

**1**

把你的想法记录在便笺上
Record your thoughts on the notes

**2**

把便笺粘在书中标记的位置
Stick the sticky note to the position marked in the book

**3**

当需要想法时，快速索引找到相应内容
Quickly find the idea when you need it

# 目录 | TABLE OF CONTENTS

| | |
|---|---|
| 阵列 | array 48 |
| 围罩 | cover 46 |
| 拼合 | split 44 |
| 泡泡 | bubble 42 |
| 拟态 | mimesis 40 |
| 螺旋 | spiral 38 |
| 剪折 | cutting and floding 36 |
| 干扰 | interfere 34 |
| 翻起 | flip 32 |
| 波折 | twists and turns 30 |

## 前言 PREFACES    面 SURFACE

## 体 VOLID    线 LINE

| | | | |
|---|---|---|---|
| 堆叠 | stacking 2 | 叠加 | superposition 52 |
| 开洞 | cave 4 | 放射 | emit 54 |
| 旋转 | rotate 6 | 渐变 | shade 56 |
| 起折 | flod 8 | 交错 | staggered 58 |
| 回折 | flodback 10 | 阵列 | array 60 |
| 螺旋 | spiral 12 | 支撑 | support 62 |
| 重复 | duplicate 14 | | |
| 聚合 | converge 16 | | |
| 削切 | cut 18 | | |
| 随机 | random 20 | | |
| 拟态 | mimicry 22 | | |
| 退叠 | stack-up 24 | | |
| 弯曲 | bend 26 | | |

| 互联 | interconnection 84 |
|---|---|
| 限定 | restric 82 |
| 围合 | enclosing 80 |
| 模糊 | blur 78 |
| 流动 | flowing 76 |
| 挤压 | extrusion 74 |
| 对比 | contrast 72 |
| 起伏 | fluctuation 70 |
| 起翘 | tinting 68 |
| 切割 | cut 66 |

## 边界 | BOUNDARY

| 作者 | author 104 |
|---|---|
| 引用 | references 102 |

## 附录 | EPILOGUE

## 介入 | INTERVENTION

| 地形 | landform 88 |
|---|---|
| 流线 | streamline 90 |
| 日照 | sunshine 92 |
| 视线 | sight 94 |
| 通风 | aeration 96 |
| 植物 | plant 98 |

# 体 / 堆叠
## VOLID / STACKING

堆叠 / stacking

### 引注 / notes

1. PAOLO VENTURELLA
2. ARCHITECTS ORANGE
3. OLE SCHEEREN，OMA

体块堆叠形成错落咬合的多样室外共享空间。
体块的进退形成丰富的室外活动场所。

The block stacks form a variety of out-of-space shared spaces that are mis-occluded.
The formation of a large body of advancing and retiring outdoor activities.

手记区 / mark

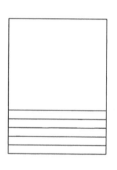
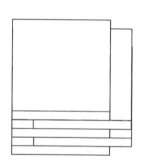

date

# 图解 / diagram

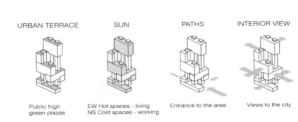

URBAN TERRACE — Public high green plazas
SUN — EW Hot spaces - living / NS Cold spaces - working
PATHS — Entrance to the area
INTERIOR VIEW — Views to the city

2011 Madrid, Spain
MIXED-USE-TOWER, BY PAOLOVENTURELLA

2011 Beirut, Lebanon
立方大厦, BY ARCHITECTS ORANGE

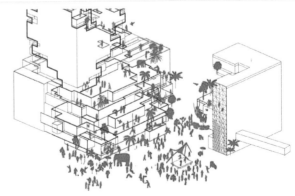

2009 Bangkok, Thailand
MAHANAKHON, BY OLE SCHEEREN, OMA

# 体 / 开洞

**VOLID / CAVE**

开洞 / cave

引注 / **notes**

1. BJARKE INGELS GROUP
2. NL ARCHITECTS
3. STEVEN HOLL

开洞形成空享模式的负空间。
环境渗透到内部空间形成对话。
体块中"洞穴与光"构成空间中的时间要素。

The open space forms the negative space of the empty mode.
The environment penetrates into the internal space to form a dialogue.
The "cave and light" in the block forms the time element in the space.

 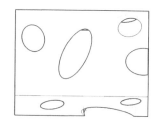

手记区 / mark

date

# 图解 / diagram

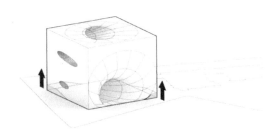

2011 Taipei, China
TEK BUILDING, BY BIG

2014 Taipei, China
台北表演艺术中心竞赛方案, BY NL ARCHITECTS

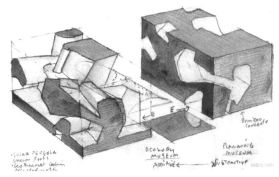

2013 Tianjing, China
天津生态城生态规划博物馆, BY STEVEN HOLL

# 体 / 旋转
VOLID / ROTATE

旋转 / rotate

## 引注 / notes

1. PAOLO VENTURELLA
2. 3XN ARCHITECTS
3. PAOLO VENTURELLA

体块通过旋转，适应场地条件的解构方式。
旋转的退层形成多样的室外空间。
旋转的方式形成场地连续性的转译。

Body block by rotating to adapt to site conditions of deconstruction.
The rotating decks form a variety of outdoor spaces.
Rotation of the formation of the continuity of the site translation.

手记区 / mark

date

图解 / diagram

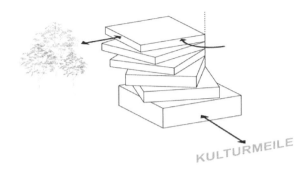

2011 Stuttgart, Germany
JOHN CRANKO BALLET SCHOOL, BY PAOLO VENTURELLA

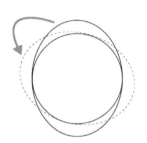

2015 Copenhagen, Denmark
NORDHAVNEN RESIDENCES, BY 3XN ARCHITECTS

2014 Unknown location
旋转像素办公楼, BY PAOLO VENTURELLA

# 体 / 起折

VOLID / FLOD

起折 / flod

## 引注 / notes

1. CEBRA
2. BJARKE INGELS GROUP
3. JDS ARCHITECTS

体块通过起折,在水平方向上呼应场地形成连续性。
在垂直方向上,丰富空间建构的多样性。
起折的空间创造通风遮阳条件。

Body block through the fold, in the horizontal direction of the formation of continuity to echo the site.
In the vertical direction, the rich diversity of space construction.
From the space to create ventilation and shade conditions.

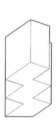
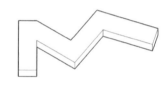

手记区 / mark

date

图解 / diagram

2015 Irkutsk, Russia
伊尔库茨克的校园设计竞赛，BY CEBRA

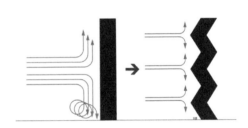

2014 Rddoure, Denmark
RODOVRE TOWER, BY BIG

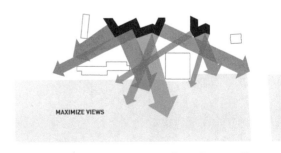

2004 Copenhagen, Denmark
HAMBROGSGADE OFFICE BUILDING, BY JDS

# 体 / 回折
## VOLID / FLODBACK

回折 / flodback

引注 / notes

1. BJARKE INGELS GROUP

2. BJARKE INGELS GROUP

3. 华黎

回折适应不同的采光要求。
回折创造相对独立的视野。
回折建构空间转折的趣味性。

Foldback to adapt to different lighting requirements.
Foldback to create a relatively independent perspective.
Foldback constructs the interest of space transition.

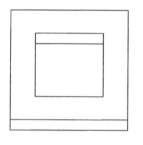 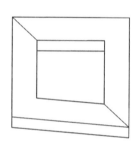

手记区 / mark

date

图解 / diagram

2013 Copenhrgen, Denmark
8TRALLET, BY BIG

2013 Holbrk Denmark
DOUBLE PERIMETER BLOCK, BY BIG

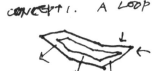

2010 Yancheng, China
水边会所, BY 华黎

# 体 / 螺旋
VOLID / SPIRAL

螺旋 / spiral

### 引注 / notes

1. BJARKE INGELS GROUP

2. HUGON KOWALSKI, ADAM WIERCINSKI, BORYS WRZESZCZ

3. 中村浩美

体块通过螺旋形成连续上升的流线。
闭合的螺旋空间形成连续的自上而下的动线。

The body block forms a continuously rising streamline by the spiral.
The closed helical space forms a continuous top-down motion line.

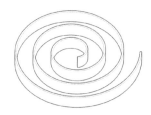

### 手记区 / mark

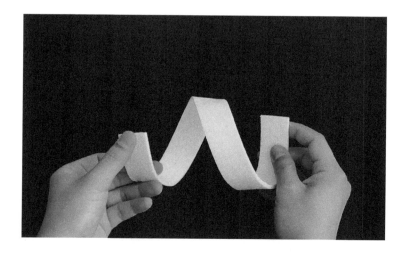

date

图解 / diagram

2010 Shanghai，China
上海世博会丹麦馆，BY BIG

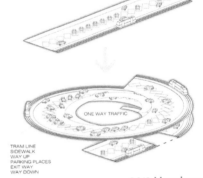

2012 Hongkong，China
空中街道，BY HUGON KOWALSKI，
ADAM WIERCINSKI，BORYS WRZESZCZ

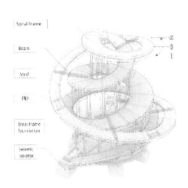

2013 Hiroshima，Japan
丝带教堂，BY 中村浩美

# 体 / 重复

VOLID / DUPLICATE

重复 / duplicate

引注 / notes

1. OFIS

2. BJARKE INGELS GROUP

3. BJARKE INGELS GROUP

单元通过给定的秩序重复建构。
重复的转译形成建构方式。
改变单元组合的秩序。

The units are repetitively constructed in a given order.
Duplicate translations form constructs.
Change the order of the unit combination.

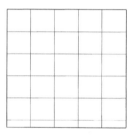
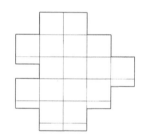

手记区 / mark

date

图解 / diagram

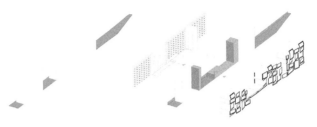

2013 Paris, France
BASKET APARTMENTS, BY OFIS

2012 Copenhrgen, DK
BORGERGADE HOUSING & PARKING, BY BIG

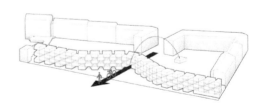

2011 Copenhrgen, DK
FREDERIKSBDRGVEJ 73, BY BIG

# 体 / 聚合

聚合 / converge

VOLID / CONVERGE

引注 / notes

1. OFIS

2. STEVEN HOLL

3. 朱亦民

向心的聚合空间与环境形成相互渗透的关系。
聚合的空间可以形成独立的视线。

To the heart of the polymerization space and the environment to form a mutual penetration relationship.
The aggregated spaces can form separate lines of sight.

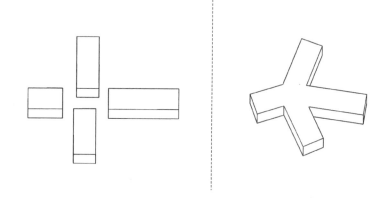

手记区 / mark

date

图解 / diagram

2005 Helsingdr, DK
PSYCHIATRIC HOSPITAL, BY OFIS

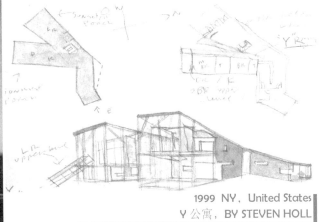

1999 NY, United States
Y 公寓, BY STEVEN HOLL

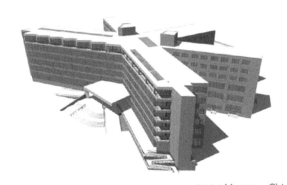

2004 Henan, China
火炬大厦, BY 朱亦民

# 体 / 削切

## VOLID / CUT

削切 / cut

### 引注 / notes

1. JDS ARCHITECTS
2. BJARKE INGELS GROUP
3. JDS ARCHITECTS

负形削切形成多样的共享空间及灰空间。
外部削切形成适应场地日照条件的空间体系。
组团削切形成各自独立的视线。

Negative cutting to form a variety of shared space and gray space.
External cutting to form a space system to adapt to the site sunshine conditions.
Group cut to form a separate line of sight.

 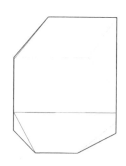

手记区 / mark

date

图解 / diagram

2008 Wroclaw, Poland
WARSAW MUSEUM OF MODERN ART, BY JDS

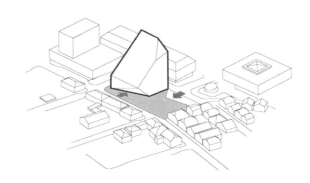

2010 Strurnger, NO
STAVANGER HOTEL, BY BIG

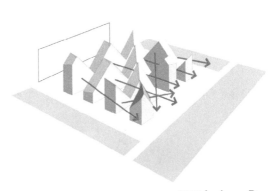

2013 Aarhus, Denmark
TAD ICEBERG, BY JDS

# 体 / 随机

VOLID / RANDOM

随机 / random

## 引注 / notes

1. SANNA

2. BJARKE INGELS GROUP

3. YOUNG & AYATA

随机组合形成多样连续空间模式。
通过错缝拼接、连续拼接，将单元进行组合，并形成自由的动线。

Random combination to form a variety of continuous spatial pattern.
By stitching splicing, continuous splicing, the combination of units, and the formation of free moving line.

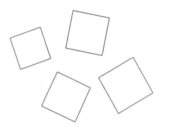

## 手记区 / mark

date

图解 / diagram

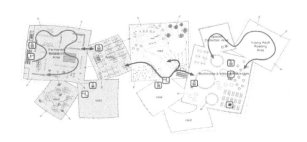

2013 Taipei, China
台中城市文化中心，BY SANNA

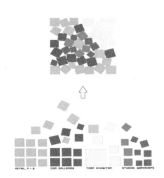

2010 Rbudhrbi, RE
ART SOUQ, BY BIG

2015 Berlin, Germany
包豪斯博物馆，BY YOUNG & AYATA

# 体 / 拟态

VOLID / MIMICRY

拟态 / Mimicry

引注 / notes

1. BJARKE INGELS GROUP
2. BJARKE INGELS GROUP
3. 宽璐建筑

通过"地景化"处理营造独立空间单元。
形成共享开放空间。
起伏的地景单元与场地自适应连接。

Through the "landscape" treatment to create an independent space unit.
Forming a shared open space.
The undulating landscape units are self-adapting to the site.

 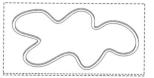

手记区 / mark

date

图解 / diagram

2010 Anywhere
DOLPHINARIUM AND WELLNESSCENTER, BY BIG

2013 Mrlmq, SE
WORLD VILLAGE OF WOMEN SPORTS, BY BIG

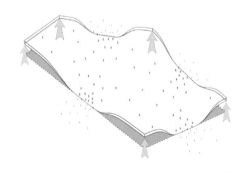

2014 Erduosi, China
鄂尔多斯多克市民中心, BY 宽璐建筑

# 体 / 退叠

## VOLID / STACK-UP

退叠 / stack-up

### 引注 / notes

1. ONAT ÖKTEM，ZEYNEP ÖKTEM
2. OPEN ARCHITECTURE
3. MELIKE ALTINIŞIK + GÜL ERTEKIN

体块通过退叠适应场地坡地条件。
退叠生成可达的丰富的户外空间。
连续的退叠形成自然场地的天际线。

Body block through the back slope to adapt to the site conditions.
Backfilling creates a rich、outdoor space.
A continuous cascade forms the skyline of a natural site.

手记区 / mark

date

图解 / diagram

2010 Ankara, Turkey
科技园绿色建筑展示与企业中心，BY ONAT ÖKTEM, ZEYNEP ÖKTEM

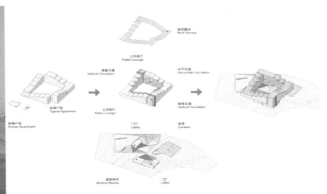

2014 Changle, China
退台方院，BY OPEN ARCHITECTURE

2015 Istanbul, Turkey
伊斯坦布尔教堂和文化中心，BY MELIKE ALTINIŞIK + GÜL ERTEKIN

# 体 / 弯曲

VOLID / BEND

弯曲 / bend

## 引注 / notes

1. BJARKE INGELS GROUP
2. BJARKE INGELS GROUP
3. AGENCE RVA

弯曲的方式能形成最大连续流线与界面。
弯曲建构体块形成互融交错的景观视野。
弯曲形成自由的外部空间边界。

Bending the way to form the largest continuous flow line and interface.
Curved construction blocks form a landscape of interlocking landscape.
Bend forms a free outer space boundary.

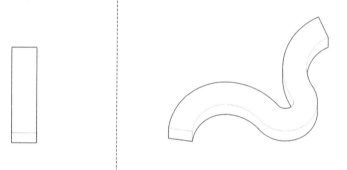

## 手记区 / mark

date

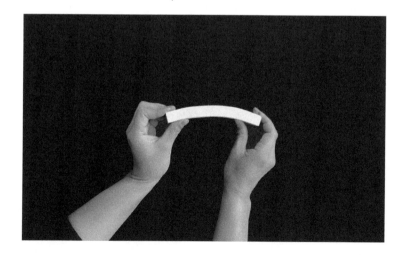

图解 / diagram

2012 Deretrden, DK
CARPARK CARSALE DWELLINGS, BY BIG

2013 Prrgue, CZ
NATIONAL LIBRARY OF THE CZECH REPUBLIC, BY BIG

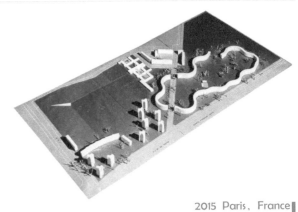

2015 Paris, France
巴黎蛇形公寓改造项目, BY AGENCE RVA

# 面 / 波折
## SURFACE / TWISTS AND TURNS

**波折 / twists and turns**

### 引注 / notes

1. 王澍
2. KENGO KUMA & ASSOCIATES
3. NORIHIKO DAN AND ASSOCIATES

连续的折坡形成大空间的覆盖。
波折增加了单元坡顶组合的多样性。

Continuous folding creates large space coverage.
The twists and turns increase the diversity of the unit tops.

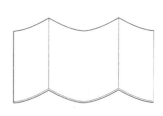

手记区 / mark

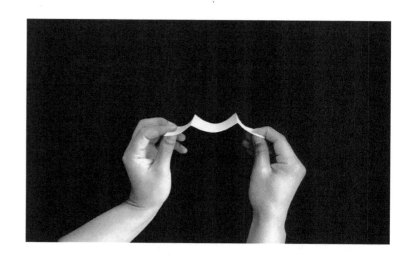

date

图解 / diagram

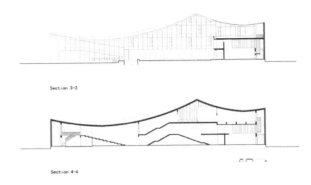

2012 Hangzhou, China
中国美术学院象山校区，BY 王澍

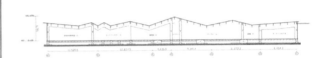

2015 Aomori, Japan
十和田市社区广场，BY KENGO KUMA & ASSOCIATES

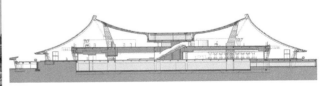

2013 Taoyuan, China
桃园国际机场，BY NORIHIKO DAN AND ASSOCIATES

# 面 / 翻起
## SURFACE / FLIP

翻起 / flip

引注 / notes

1. TEAM730
2. 西泽立卫
3. WE ARCHITECTURE

大片连续的翻起的面形成节奏空间。
通过场地覆盖，形成与环境适应的空间。
渐变翻起的表皮生成标识性空间。

Large continuous flip surface to form a rhythm space.
Through the site coverage, and the environment to adapt to the formation of space.
Gradient of the epidermis from the identification of space.

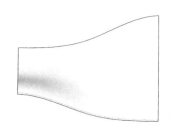

手记区 / mark

date

图解 / diagram

**2015 Lushan，China**
庐山世界建筑博览园多功能街区，**BY TEAM730**

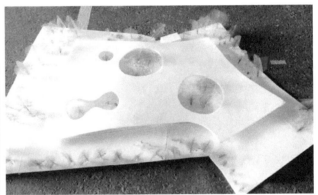

**2014 Karuizawa，Japan**
千住博博物馆，**BY** 西泽立卫

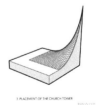

**2011 Vaaler，Norway**
挪威瓦勒教堂，**BY WE ARCHITECTURE**

# 面 / 干扰

SURFACE/INTERFERE

干扰 / interfere

引注 / notes

1. *OOZN
2. AEDAS
3. IÑAKI ECHEVERRIA

干扰将动态基因引入表皮生成当中。
动态的流线被干扰后作为建筑表皮生成基因。
扰动的表皮形成连续的空间界面。

Interference will be the introduction of dynamic genes into the epidermis.
Dynamic streamlines are interfered with as building envelope genes.
The perturbed epidermis forms a continuous spatial interface.

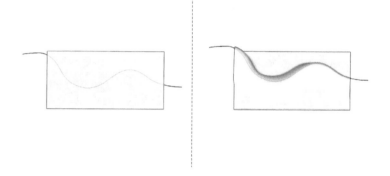

手记区 / mark

date

图解 / diagram

2015 Bangsar，Malaysia
BUKIT PANTAI RESIDENCE，BY +OOZN

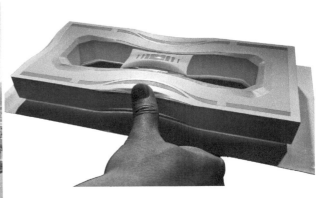

2011 Beijing，China
北京新浪总部大楼，BY AEDAS

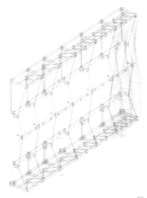

2014 Tabasco，Mexico
墨西哥利物浦百货大楼，BY IÑAKI ECHEVERRIA

# 面 / 剪折
## SURFACE / CUTTING AND FLODING

剪折 / cutting and floding

引注 / notes

1. **CROSSBOUNDARIES**
2. **KADAWITTFELDARCHITEKTUR**
3. 标准营造

以剪折表皮包络生成内部空间界面。
面片折起生成连续的坡道空间。
通过剪折折叠形成流动空间。

The internal space interface is generated by cutting and folding the skin envelope.
The dough sheet is folded to create a continuous ramp space.
The flow space is formed by shearing and folding.

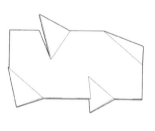

手记区 / mark

date

图解 / diagram

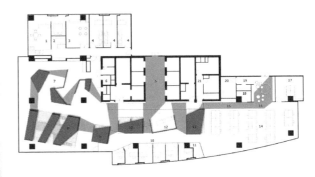

2012 Beijing, China
ELITE GROUP HEADQUARTERS, BY CROSSBOUNDARIES

2006 Salzburg, Austria
PAPPAS HEADQUARTERS, BY KADAWITTFELDARCHITEKTUR

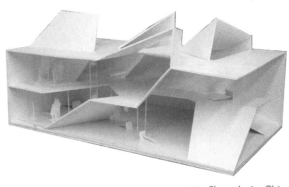

2011 Shanghai, China
折叠的房子, BY 标准营造

# 面 / 螺旋
SURFACE / SPIRAL

螺旋 / spiral

引注 / notes

1. 藤本壮介
2. FRANEK-ARCHITECTS
3. HERZOG & DE MEURON

螺旋形成多种交互动线，弱化层与层之间的关系。
形成连续回转的空间动线。
建构形成开放上升的空间体验。

The spiral forms a variety of interacting lines, weakening the relationship between layers and layers.
The formation of continuous rotation of the space line.
Building an open space experience.

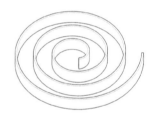

手记区 / mark

date

图解 / diagram

2012 Belgrade, Serbia
BETON HALA 滨水中心，BY 藤本壮介

2015 Chata Slamenka, Czech Republic
SKYWALK, BY FRANEK-ARCHITECTS

2015 Oxfordshire, United Kingdom
牛津郡政府学校，BY HERZOG & DE MEURON

# 面 / 拟态

SURFACE / MIMESIS

拟态 / mimesis

引注 / notes

1. SANAA

2. BOHLIN CYWINSKI JACKSON

3. ERGINOĞLU + ÇALIŞLAR ARCHITECTS

通过自然的抽象空间建立环境的内外互动关系。
空间以水平延展的姿态介入环境，形成多样空间层次。

Through the nature of the abstract space to establish the internal and external environment interaction.
Space to horizontal extension of the attitude involved in the environment, the formation of a variety of spatial levels.

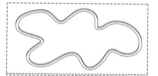

手记区 / mark

date

图解 / diagram

2009 London，UK
蛇形画廊铝板展区，BY SANAA

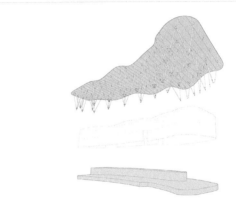

2015 Qingdao，China
青岛明珠游客中心，BY BOHLIN CYWINSKI JACKSON

2008 Amburan，Indonesia
AMBURAN 海滩，BY ERGINOĞLU + ÇALIŞLAR ARCHITECTS

# 面 / 泡泡

SURFACE / BUBBLE

泡泡 / bubble

## 引注 / notes

1. SELGAS CANO
2. 马岩松
3. MVRDV

包裹的表皮柔滑过渡形成连续的空间。
泡泡作为单元连续拼接形成群体空间。

The smooth transition of the skin of the package forms a continuous space.
Bubble as a unit to form a group of continuous splicing space.

手记区 / mark

date

图解 / diagram

2015 London, UK
蛇形展馆, BY SELGAS CANO

2012 Beijing, China
胡同泡泡 32 号, BY 马岩松

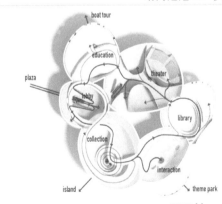

2013 Hangzhou, China
杭州动漫博物馆, BY MVRDV

43

# 面 / 拼合

SURFACE / SPLIT

拼合 / split

### 引注 / notes

1. AL_A
2. PLAN B ARCHITECTS + JPRCR ARCHITECTS
3. 让·努维尔

通过拼合形成有节奏的韵律空间，建立空间仪式感。
自由拼合，自由扩散，与自然融合。
随机拼合形成独特的空间模式。

Through the formation of a rhythm of space，the establishment of a sense of space ritual.
Free integration，free diffusion，and natural fusion.
Random combination of the formation of a unique spatial pattern.

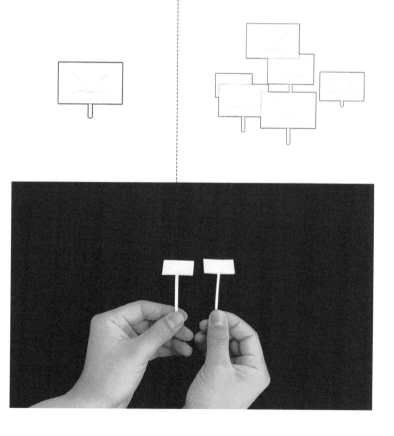

### 手记区 / mark

date

图解 / diagram

2015 Abu Dhabi
ABU DHABI MOSQUE, BY AL_A

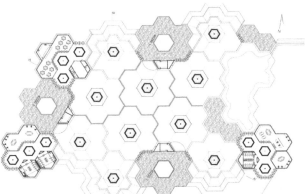

2008 Antioquia, Colombia
ORQUIDEORAMA, BY PLAN B ARCHITECTS + JPRCR ARCHITECTS

2010 Qatar
卡塔尔国家博物馆, BY 让·努维尔

# 面 / 围罩

SURFACE/COVER

围罩 / cover

引注 / notes

1. NOIZ ARCHITECTS + BIO ARCHITECTURE

2. SSD ARCHITECTURE

3. 马岩松（MAD）

通过围罩形成围护表皮，作为建筑遮阳通风的功能。
将分散的空间化整为零，形成统一完整的空间。
整合场地条件形成独特的界面，整合内部空间。

Through the formation of enclosure enveloping epidermis, as a building shade ventilation function.
Will be dispersed into a zero space, forming a unified and complete space.
Integration of site conditions to form a unique interface, integration of internal space.

 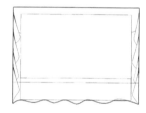

手记区 / mark

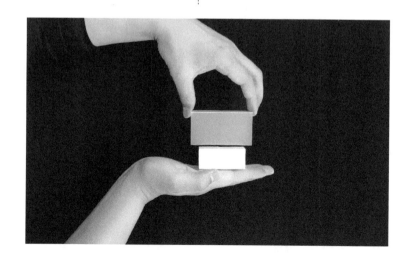

date

图解 / diagram

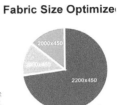

2010 Nantou, China
南投工业科技研究所，BY NOIZ ARCHITECTS + BIO

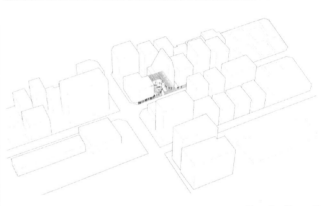

2011 Seoul, Korea
松坡区某小型住宅，BY SSD ARCHITECTURE

2015 Aichi, Japan
叶草之家幼儿园，BY 马岩松（MAD）

# 面 / 阵列
## SURFACE / ARRAY

阵列 / array

### 引注 / notes

1. VO TRONG NGHIA ARCHITECTS
2. 李兴刚
3. 董功（直向建筑）

竖向上通过节奏形成空间边界。
水平上通过阵列放样空间。

Vertical through the rhythm of the formation of spatial boundaries.
Horizontally through the array lofting space.

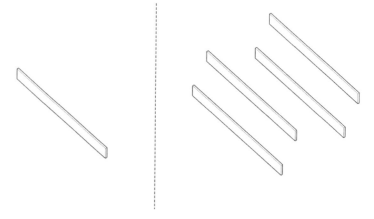

手记区 / mark

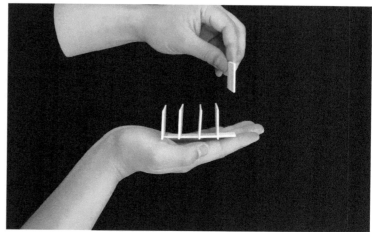

date

图解 / diagram

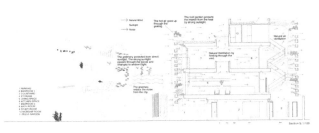

2012 Saigon, Vietnam
STACKING GREEN, BY VO TRONG NGHIA ARCHITECTS

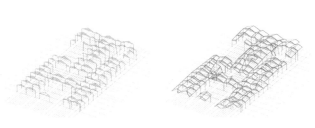

2013 Jixi, China
绩溪博物馆, BY 李兴刚

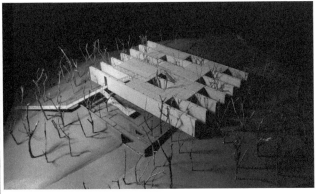

2013 Jinan, China
小清河湿地展馆, BY 董功（直向建筑）

49

# 线 / 叠加
LINE / SUPERPOSITION

叠加 / superposition

## 引注 / notes

1. 隈研吾
2. 隈研吾
3. IMPROMPTU PROJECT

分层叠加形成支撑体系。
分层叠加形成跨度空间。
分层叠加建构组合空间。

Hierarchical Superposition to form a support system.
Hierarchical Superposition forms a span space.
Hierarchical Superposition constructs combination space.

手记区 / mark

date

图解 / diagram

2011 Toyama, Japan
CAFÉ KUREON, BY 隈研吾

2007 London, UK
SAKE NO HANA, BY 隈研吾

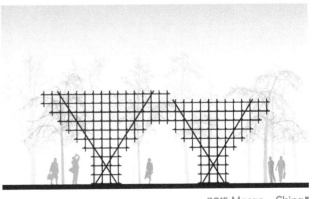

2015 Macao, China
澳门三胞树竹装置, BY IMPROMPTU PROJECT

# 线 / 放射

LINE / EMIT

### 引注 / notes

1. LUX NOVA BY EASTON+COMBS
2. 托马斯·西斯维克
3. MOXON ARCHITECTS

手记区 / mark

date

放射 / emit

放射的结构自由组合形成空间。
放射后的表皮生成生动自然的界面。

The radiating structures are freely combined to form spaces.
Radiation after the epidermis vivid and natural interface.

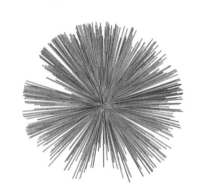

图解 / diagram

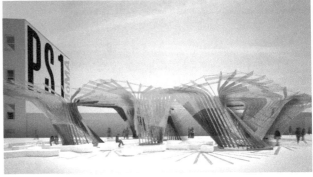

2010 TNew York, United States
P.S.1 2010 ENTRY, BY LUX NOVA BY EASTON+COMBS

2010 Shanghai, China
上海世博会英国馆, BY 托马斯·西斯维克

2010 England, UK
"刺猬建筑", BY MOXON ARCHITECTS

# 线 / 渐变
LINE / SHADE

渐变 / shade

### 引注 / notes

1. ADARC ASSOCIATES

2. PIER ALESSIO RIZZARDI

3. nARCHITECTS

渐变的节奏建构空间的骨架系统。
变异的重复单元成为空间的生成逻辑。

Gradient rhythm constructs the skeleton system of the space. The mutated repeating unit becomes the generating logic of the space.

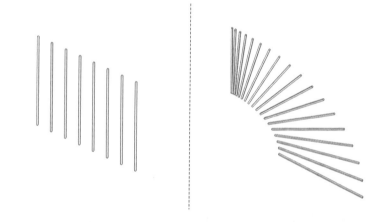

手记区 / mark

date

图解 / diagram

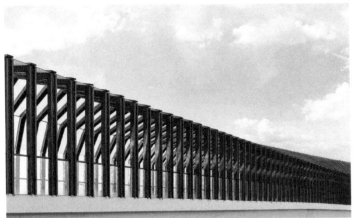

2015 Foshan, China
佛山新城天桥，BY ADARC ASSOCIATES

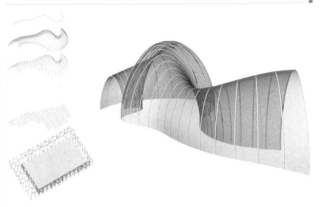

2014 Venice, Italy
PARASITE PAVILION, BY PIER ALESSIO RIZZARDI

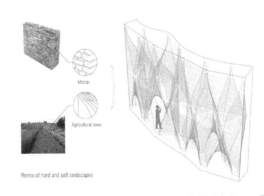

2008 Léren, France
WINDSHAPE, BY nARCHITECTS

# 线 / 交错

LINE / STAGGERED

交错 / staggered

## 引注 / notes

1. HADUWA ARTS & CULTURE INSTITUTE
2. 3XN ARCHITECTS
3. ANDRÉS JAQUE

交错建立单层网架系统，形成空间的覆盖。
交错的线条建立表皮系统。
交错张拉线条将柔性杆件转换为刚度空间。

Staggered to establish a single-layer network system, the formation of space coverage.
Staggered lines build up the epidermal system.
Staggered lines convert flexible rods into rigid spaces.

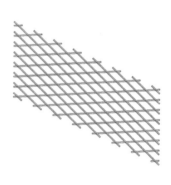

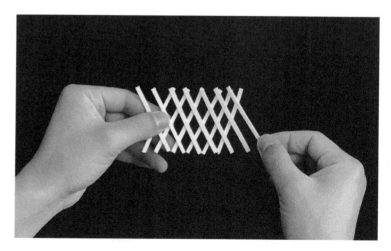

手记区 / mark

date

图解 / diagram

2015 Apam, Ghana
PAVILION, BY HADUWA ARTS & CULTURE INSTITUTE

2006 Berlin, Germany
LEHRTER STADTQUARTIER, BY 3XN ARCHITECTS

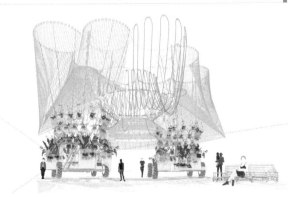

2015 Long Island City, United States
MOMA PS1 YAP 2015 - COSMO, BY ANDRÉS JAQUE

59

# 线 / 阵列
## LINE / ARRAY

阵列 / array

引注 / notes

1. ARCVS

2. MIA DESIGN STUDIO

3. PIUARCH

重复性阵列形成空间场所。
密集的线性阵列形成空透界面。

Repetitive arrays form spatial locations.
The dense linear array forms a vacant interface.

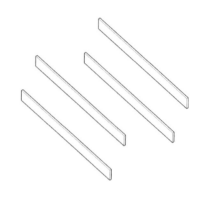

手记区 / mark

date

图解 / diagram

2015 Hungary
匈牙利音乐之家，BY ARCVS

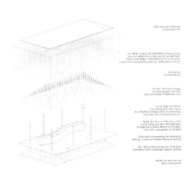

2015 Hanoi, Vietnam
VIETNAMESE FOOD PAVILION, BY MIA DESIGN STUDIO

2015 Milan, Italy
意大利国家电力馆，BY PIUARCH

# 线 / 支撑
LINE / SUPPORT

支撑 / support

## 引注 / notes

1. ATELIER YOKYOK

2. SERIE ARCHITECTS

3. GUILLAUME MAZARS

模块化的支撑连接形成自由空间。
骨架式支撑形成空透的表皮系统。

The modular support connections form a free space.
Skeletal support to form a transparent skin system.

手记区 / mark

date

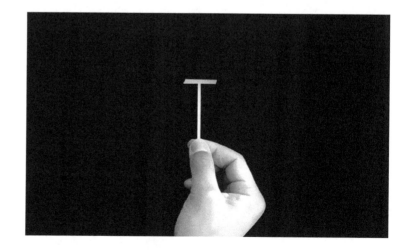

图解 / diagram

2015 Budapest, Hungary
"树的自由", BY ATELIER YOKYOK

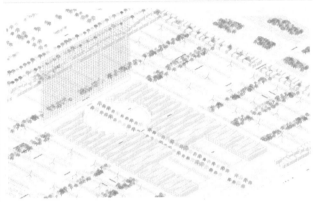

2013 Kazakhstan, Astana
阿斯塔纳世博会竞赛方案, BY SERIE ARCHITECTS

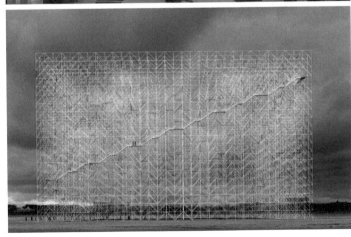

2015 Novossibirsk, Russia
REVEAL THE ABSENCE, BY GUILLAUME MAZARS

# 边界 / 切割
## BOUNDARY / CUT

切割 / cut

### 引注 / notes

1. LINÉAIRE A
2. VAUMM
3. TOYO ITO

以场地为条件，通过日照、动线等作为空间边界切割逻辑。
基于天际线关系对边界切割，与环境建立联系。

The conditions for the site, through the sun, moving lines, etc. as a boundary cutting logic space.
Boundary cutting based on the skyline relationship, and establish contact with the environment.

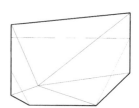

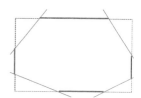

手记区 / mark

date

图解 / diagram

2012 Stockholm, SWE
ODENPLAN ATRIUM, BY LINÉAIRE A

2015 Gipuzkoa, Spain
PARK & BASQUE PELOTA COURT, BY VAUMM

2014 LocationTokyo, Japan
ZA KOENJI PUBLIC THEATRE, BY TOYO ITO

# 边界 / 起翘
## BOUNDARY / TINTING

起翘 / tinting

引注 / notes

1. CASTAÑEDA
2. NEW WAVE ARCHITECTURE
3. STUDIO O + A

延续场地条件，起翘形成起伏的边界。
转换场地边界，边界形成复合的动线。

Continuation of site conditions, upturned the formation of ups and downs of the border.
Conversion site boundary, the formation of a complex boundary line.

手记区 / mark

date

图解 / diagram

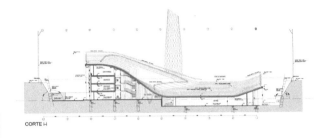

2014 Cordoba, Argentina
CÓRDOBA CULTURAL CENTER, BY CASTAÑEDA

2015 Tehran, Iran
SUSTAINABLE OFFICE BUILDING, BY NEW WAVE ARCHITECTURE

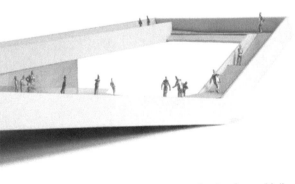

2010 Amsterdam, Holland
URBAN BEACH, BY STUDIO O + A

# 边界 / 起伏
## BOUNDARY / FLUCTUATION

起伏 / fluctuation

### 引注 / notes

1. JDS ARCHITECTS
2. JEAN-PHILIPPE PARGADE
3. ZAHA HADID ARCHITECTS

以坡地为线索建构起伏的自然边界。
通过连续起伏的坡面建构空间。

Based on the slope as the clue, construct the natural boundary of fluctuation.
Constructing space through continuous undulating slopes.

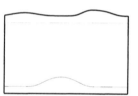

手记区 / mark

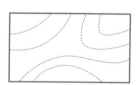

date

图解 / diagram

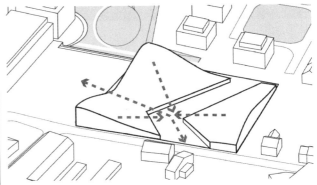

2008 Aalter, Belgium
AAL DE WARANDE IN AALTER, BY JDS

2014 Paris, France
巴黎校园，BY JEAN-PHILIPPE PARGADE

2013 Baku, Azerbaijan
阿利耶夫文化中心，BY ZAHA HADID ARCHITECTS

# 边界 / 对比
## BOUNDARY / CONTRAST

对比 / contrast

### 引注 / notes

1. SET ARCHITECTS

2. SKOPE

3. YOON SPACE

利用形态对比建构空间关系。
利用材质对比建构空间关系。
利用界面对比形成建构体系。

Construction of spatial relations by morphological contrast.
Constructing spatial relationship by material contrast.
Using the interface contrast to form the construction system.

手记区 / mark

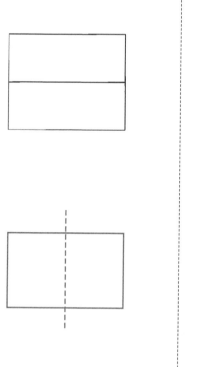

date

图解 / diagram

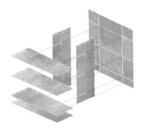

2016 Bologna, Italy
BOLOGNA SHOAH MEMORIAL, BY SET ARCHITECTS

2013 Brussels, Belgium
议会法语部办公楼, BY SKOPE

2015 Korea
陶艺家 JUNG GIL-YOUNG 展览馆, BY YOON SPACE

# 边界 / 挤压
## BOUNDARY / EXTRUSION

挤压 / extrusion

引注 / notes

1. NIMA NIAN + BEHDAD HEYDARI

2. 3XN

3. THOMAS PHIFER + PARTNERS

以场地渗透为条件，挤压生成空间边界。
以场地连接为条件，挤压形成界面空间。

On the condition of site infiltration, squeezing generates spatial boundaries.
On the condition of site connection, the interface space is formed by extrusion.

手记区 / mark

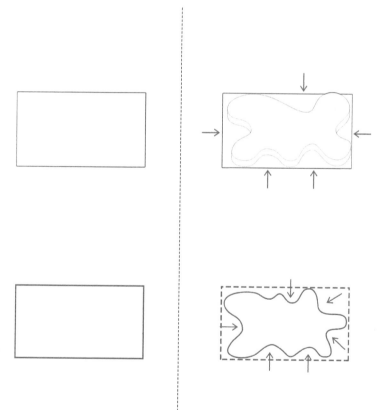

date

图解 / diagram

2016 Krakow, Poland
OXYGEN HOME, BY NIMA NIAN BEHDAD HEYDARI

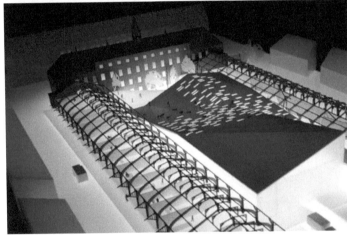

2015 Orr Hus, Denmark
火车站文化中心，BY 3XN

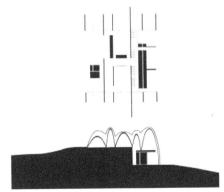

2013 Long Island, United States
长岛住宅，BY THOMAS PHIFER + PARTNERS

75

# 边界 / 流动
## BOUNDARY / FLOWING

流动 / flowing

### 引注 / notes

1. BJARKE INGELS GROUP

2. 3XN

3. SANAA

以地形为线索生成流动的界面。
流动的界面以薄壳结构建构形成独立的支撑体系。

Creates a flowing interface with the terrain as a clue.
The flow interface is constructed in a thin shell structure to form an independent support system.

手记区 /mark

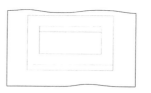 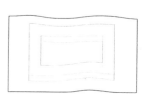

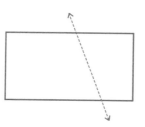 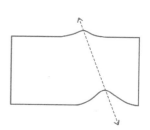

date

图解 / diagram

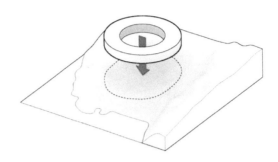

2013 Nuuk, RJD
GREENLANDS NATIONAL GALLERY OF ART, BY BIG

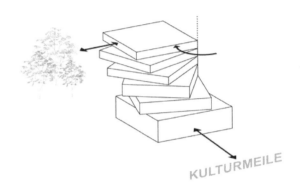

KULTURMEILE

2013 Yong-in Si, South Korea
TRICIRCLE SOUTH KOREA, BY 3XN

2011 Lausanne Confederation, Switzerland
瑞士劳力士学习中心, BY SANAA

# 边界 / 模糊

## BOUNDARY / BLUR

模糊 / blur

### 引注 / notes

1. CONSTRUCTO CHILE
2. UNK PROJECT
3. CHEUNGVOGL

空间自然凹入弱化界面对场地的分隔。
利用视觉模糊形成环境互融的界面。

Space naturally concave, weakening the interface of the separation of the site.
The use of visual fuzzy environment to form the interface.

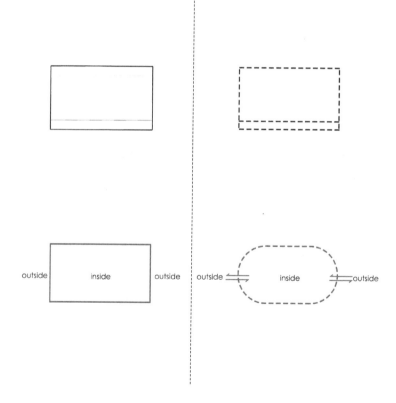

### 手记区 / mark

date

图解 / diagram

2015 Santiago，Chile
"反思"YAP 获奖方案，**BY CONSTRUCTO CHILE**

2016 Moscow，Russia
**ATOMIC ENERGY PAVILION DESIGN，BY UNK PROJECT**

2012 bei Duesseldorf，Germany
**TAGHAUS COLLECTION，BY CHEUNGVOGL**

# 边界 / 围合
## BOUNDARY / ENCLOSING

围合 / enclosing

### 引注 / notes

1. ANTONIO VIRGA ARCHITECTECTURE

2. 都市实践

3. SCENIC ARCHITECTURE

将空间作为自由边界包裹环境。
以公共空间为中心围合形成拓展的"界面"。

Space is wrapped as a free boundary environment.
Centered around the public space to form the expansion of the "interface".

手记区 / mark

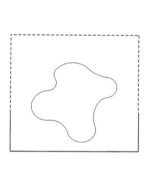
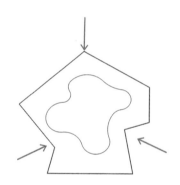
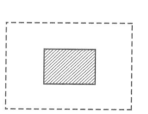
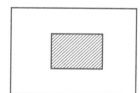

date

图解 / diagram

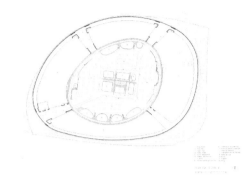

2012 Angers, France
L'ATOLL 购物中心，BY ANTONIO VIRGA ARCHITECTECTURE

2011 Shenzhen, China
美伦服务公寓，BY 都市实践

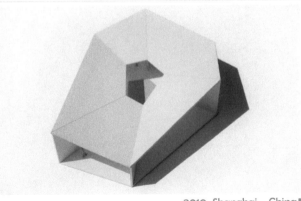

2010 Shanghai, China
金陶村村民活动室，BY SCENIC ARCHITECTURE

# 边界 / 限定
## BOUNDARY / RESTRICT

限定 / restrict

### 引注 / notes

1. HESS TALHOF KUSMIER
2. URKO SANCHEZ
3. BRUNET

自由单元组合通过边界的界定生成空间。
边界作为场地条件限定空间和环境的关系。

The combination of free units creates space by delimiting boundaries.
Boundaries define the relationship between space and environment as site conditions.

### 手记区 / mark

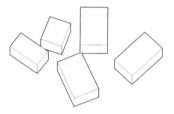 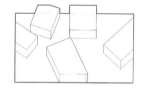

date

图解 / diagram

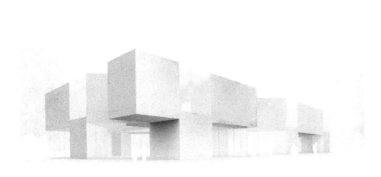
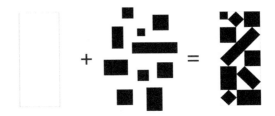

2015 Bauhaus, Germany
包豪斯博物馆竞赛，BY HESS TALHOF KUSMIER

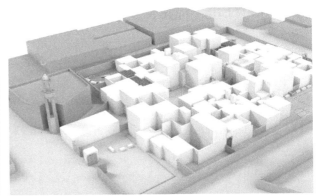

2015 Djibouti, Africa
吉布提无车村庄，BY URKO SANCHEZ

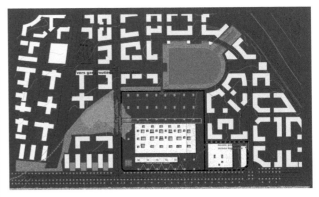

2012 Jossigny, France
CENTRAL HOSPITAL, BY BRUNET

# 边界 / 互联
## BOUNDARY/ INTERCONNECTION

互联 / interconnection

### 引注 / notes

1. 3XN
2. STEVEN HOLL
3. IARC ARCHITECTS

由基本空间通过互联形成互通的空间组合。
单元通过互联组合的方式形成整体。

From the basic space through the formation of interconnection space combination.
The units are integrated by interconnection.

手记区 / mark

 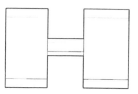

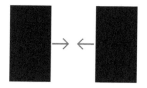 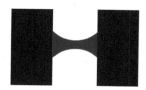

date

图解 / diagram

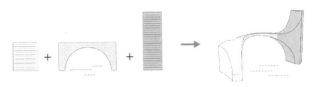

2008 Copenhagen, Denmark
CPH ARCH, BY 3XN

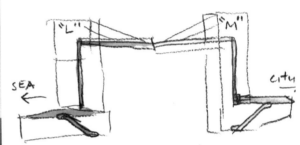

PUBLIC ESCALATORS TO SEA & CITY TERRACES
LARGE ELEVATORS TO BRIDGE

2015 Copenhagen, Denmark
LM HARBOR GATEWAY, BY STEVEN HOLL

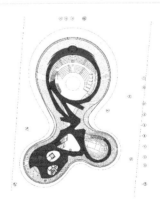

2010 Inchon, South Korea
TRI-BOWL 复合文化空间, BY IARC ARCHITECTS

85

# 介入 / 地形
## INTERVENTION / LANDFORM

地形 / landform

### 引注 / notes

1. MELIKE ALTINISIK & GUL EERTERKIN

2. ZIYA IMREN

3. BROOKS + SCARPA ARCHITECTS

场地高差作为空间建构生成的线索。
"地景"的方式形成多样的建构策略。

Site height difference as a clue to spatial construction.
"Landscape" approach to the formation of a variety of construction strategies.

手记区 / mark

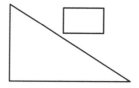 

date

图解 / diagram

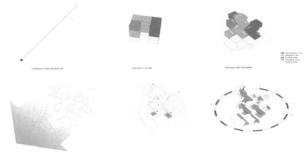

2015 Turkey, Istanbul
礼拜堂和文化中心竞赛，BY MELIKE ALTINISIK & GUL EERTERKIN

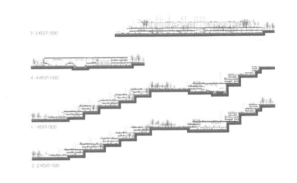

2014 Turkey, Istanbul
BEYKOZ 学校综合项目竞赛，BY ZIYA IMREN

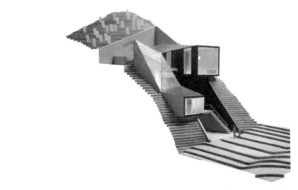

2013 Los Angeles, US
VAIL GRANT HOUSE, BY BROOKS + SCARPA ARCHITECTS

# 介入 / 流线
INTERVENTION / STREAMLINE

流线 / streamline

### 引注 / notes

1. WE ARCHITECTURE
2. JACQUES FERRIER ARCHITECTURE
3. BJARKE INGELS GROUP

流线介入空间内部分隔，形成空间建构策略。
不同的流线连接贯通，生成适应场地关联的空间。

The streamline intervenes in the interior of the space to form the space construction strategy.
Different flow lines connected through to generate space to adapt to the site associated.

手记区 / mark

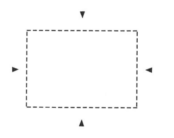
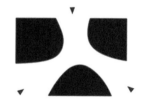

date

图解 / diagram

2010 Holte, Danmark
MARIEHØ 文化中心竞赛设计，BY WE ARCHITECTURE

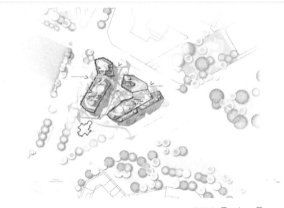

2016 Paris, France
MULTI-LAYERED CITY，BY JACQUES FERRIER ARCHITECTURE

2015 Copenhagen, Denmark
哥本哈根垃圾回收中心，BY BIG

# 介入 / 日照
## INTERVENTION / SUNSHINE

引注 / notes

1. JDS ARCHITECTS
2. JDS ARCHITECTS
3. MIBA ARCHITECTS

日照 / sunshine

基于场地日照最优化策略介入空间生成。
日照成为空间建构的动因。

Intervention space generation based on site sunlight optimization strategy.
Sunshine is the motive of space construction.

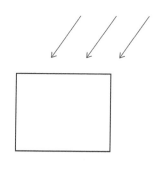
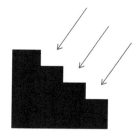
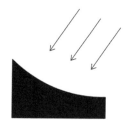
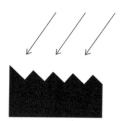

手记区 / mark

date

图解 / diagram

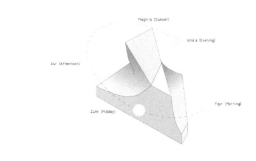

2005 Copenhagen, Denmark
BAT THE BATTERY, BY JDS

TRIANGLE

2004 Hague, Netherlands
EUROPEAN PATENT OFFICE, BY JDS

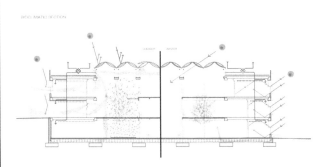

2015 Nicosia, Cyprus
UNIVERSITY OF CYPRUS MEDICAL SCHOOL, BY MIBA

# 介入 / 视线
## INTERVENTION / SIGHT

视线 / sight

### 引注 / notes

1. JDS ARCHITECTS
2. JDS ARCHITECTS
3. JDS ARCHITECTS

独立的取景形成空间生成机制。
视线作为空间生成及变形的建构策略。

Independent framing formation space generation mechanism.
Line of sight as the construction strategy of space generation and deformation.

手记区 / mark

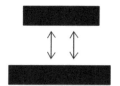 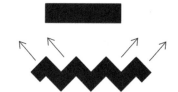

date

图解 / diagram

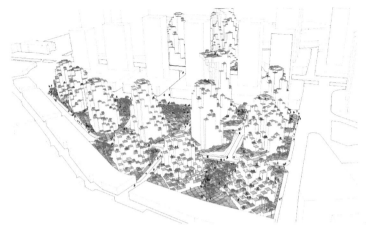

2011 Rio de Janeiro, Brazil
RIO OLYMPIC PORT, BY JDS

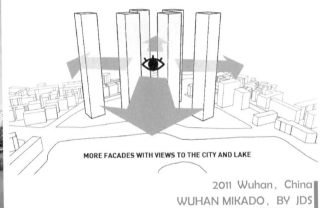

MORE FACADES WITH VIEWS TO THE CITY AND LAKE

2011 Wuhan, China
WUHAN MIKADO, BY JDS

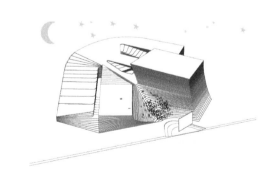

2014 Stavanger, Norway
STA STAVANGER CONCERT HALL, BY JDS

# 介入 / 通风
## INTERVENTION / AERATION

通风 / aeration

### 引注 / notes

1. AND ARCHITECTS
2. TEAM.BREATHE.AUSTRIA
3. 素朴建筑工作室

风道作为气候策略介入空间生成线索。
通风成为空间内部组织策略。

The wind tunnel as a climate strategy involves space generation clues.
Ventilation becomes a spatial internal organizational strategy.

手记区 / mark

date

图解 / diagram

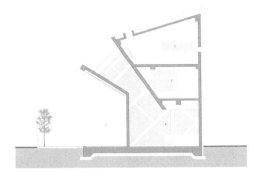
2016 Gyeonggi-do, South Korea
百叶幕墙小屋, BY AND

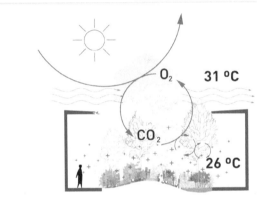
2015 Milan, Italy
米兰世博会奥地利馆, BY **TEAM.BREATHE.AUSTRIA**

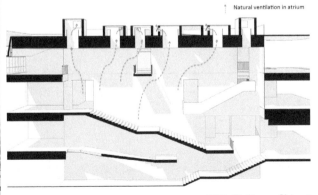
2015 Beijing, China
清华大学南区食堂及就业指导中心, BY 素朴建筑工作室

# 介入 / 植物

INTERVENTION / PLANT

植物 / plant

### 引注 / notes

1. JDS ARCHITECTS
2. 藤本壮介
3. VO TRONG NGHIA ARCHITECTS

空间以植物的介入形成互融的边界和场所。
植物表征环境形成融合的空间形态。

Space to plant the formation of inter-integration of the boundaries and places.
Plants characterize the environment to form a fused spatial form.

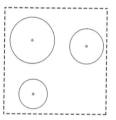

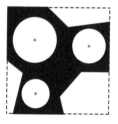
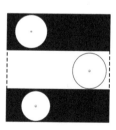

手记区 / mark

date

图解 / diagram

2011 Rio de Janeiro，Brazil
RIO OLYMPIC PORT，BY JDS

2015 Paris，France
巴黎综合理工学习中心，BY 藤本壮介

2014 Ho Chi Minh City，Vietnam
HOUSE FOR TREES，BY VO TRONG NGHIA ARCHITECTS

# 引用 | REFERENCES

书中插图分别出自于（按出现顺序）：
The illustrations in the book are from (in order of appearance):

1. Paolo Venturella Architecture；Architects Orange；OMA office
2. Bjarke Ingels Group；NL Architects；Steven Holl Architects
3. Paolo Venturella Architecture；3XN Architects；Paolo Venturella Architecture
4. CEBRA architecture；Bjarke Ingels Group；JDS Architects
5. Bjarke Ingels Group；Bjarke Ingels Group；华黎（Trace Architecture Office）
6. Bjarke Ingels Group；Hugon Kowalski，Adam Wiercinski，Borys Wrzeszcz；中村浩美
7. OFIS Architects；Bjarke Ingels Group；Bjarke Ingels Group
8. OFIS Architects；Steven Holl Architects；朱亦民
9. JDS Architects；Bjarke Ingels Group；JDS Architects
10. SANNA；Bjarke Ingels Group；Young & Ayata
11. Bjarke Ingels Group；Bjarke Ingels Group；宽璐建筑
12. Onat Oktem，Zeynep Öktem；Open Architecture；Melike Altınışık + Gül Ertekin
13. Bjarke Ingels Group；Bjarke Ingels Group；Agence RVA
14. 王澍（业余建筑工作室）；Kengo Kuma & Associates；Norihiko Dan and Associates
15. TEAM730；西泽立卫；WE Architecture
16. +OOZN design；Aedas；Iñaki Echeverria
17. Crossboundaries；Kadawittfeldarchitektur；标准营造（Standardarchitecture）
18. 藤本壮介；Franek-Architects；Herzog & de Meuron
19. SANNA；Bohlin Cywinski Jackson；ErginoGlu + Çalışlar Architects
20. Selgas Cano；马岩松（MAD）；MVRDV
21. AL_A；Plan B Architects + JPRCR Architects；让·努维尔（Jean Nouvel）
22. NOIZ architects + BIO architecture；SSD Architecture；马岩松（MAD）
23. Vo Trong Nghia Architects；李兴刚；董功（直向建筑）

24. 隈研吾；隈研吾；Impromptu Project
25. LUX NOVA by EASTON+COMBS；Thomas Heatherwick；Moxon Architects
26. ADARC Associates；Pier Alessio Rizzardi；nArchitects
27. Haduwa Arts & Culture Institute；3XN Architects；Andrés Jaque
28. ARCVS；MIA Design Studio；PIUARCH
29. Atelier YokYok；Serie Architects；Guillaume Mazars
30. Linéaire A；VAUMM；TOYO ITO
31. Castañeda；New Wave Architecture；Studio O+A
32. JDS Architects；Jean-Philippe Pargade；Zaha Hadid Architects
33. SET Architects；SKOPE；Yoon Space
34. Nima Nian + Behdad Heydari；3XN Architects；Thomas Phifer + Partners
35. Bjarke Ingels Group；3XN Architects；SANNA
36. Constructo Chile；UNK Project；Cheungvogl
37. Antonio Virga Architectecture；都市实践（URBANUS）；Scenic Architecture
38. Hess Talhof Kusmier；Urko Sanchez；BRUNET
39. 3XN Architects；Steven Holl Architects；IARC Architects
40. Melike Altinisik & Gul Eerterkin；Ziya Imren；Brooks + Scarpa Architects
41. WE Architecture；Jacques Ferrier Architecture；Bjarke Ingels Group
42. JDS Architects；JDS Architects；Miba Architects
43. JDS Architects；JDS Architects；JDS Architects
44. AND Architects；Team.Breathe.Austria；素朴建筑工作室
45. JDS Architects；藤本壮介；Vo Trong Nghia Architects

# 作者简介 | ABOUT THE AUTHOR

何东明

湖北美术学院讲师 / 栖行建筑设计事务所创始人及主持建筑师

2009 年毕业于昆明理工大学建筑历史与理论硕士专业，于 2014 年创立栖行建筑设计事务所，作品曾参加 2010CA'ASI 中国新锐建筑创作展及 2016 米兰国际设计周卫星展．

其中设计作品荣获多项国际竞赛奖：

1. 尼泊尔 - 西藏地震全球建筑师灾后重建国际建筑设计竞赛．三等奖 .2015（中国 / 尼泊尔）
2. Design a Beautiful House 国际建筑设计竞赛．三等奖 .2015（英国）
3. 2015 中国建筑学会国际青年建筑设计竞赛．提名奖
4. "Chandigarh Unbuilt Completing the Capitol" 国际建筑设计竞赛．一等奖 .201（印度）
5. "Sequoia Climbing Space" 国际建筑设计竞赛．优秀奖 .2016（西班牙）
6. 2016 上海市重大文化设施国际青年建筑师设计竞赛活动．三等奖
7. 2016 米兰设计周—中国高等院校设计学科师生优秀作品展．学院奖（中国 / 意大利）
8. 2016 重庆南滨特区超高层住宅绿色生态空间设计国际竞赛．入围奖
9. 2016 趣城计划·国际设计竞赛．优秀独立作品奖
10. 43 届日新工业株式会社建筑设计竞赛．佳作奖 .2016（日本）
11. 2017 米兰设计周—中国高等院校设计学科师生优秀作品展．学院奖（中国 / 意大利）
12. "Experiential Beer Garden" 国际建筑设计竞赛．入围奖 .2017（意大利）